S0-DVV-127

HIDDEN CREATURE FEATURES

JANE PARK

MILLBROOK PRESS / MINNEAPOLIS

Do you see our ADAPTATIONS—

a TAIL,
a CLAW,
a HORN,
or BEAK?

Others aren't so easy to spot.

CAN YOU FIND THEM?

Take a peek!

I swim for fish
by flapping my FLIPPERS.
I grab them in my bill . . .

WITH MY GRIPPERS.

PENGUINS don't have teeth. Instead, they have **BRISTLES** on their tongues to help them hold onto their food. The bristles are made of keratin—the same protein that makes up your hair and fingernails.

My RED EYES surprise—
they're so bright and bold.
But when I'm asleep . . .

THEY'RE LACY AND GOLD.

When a **RED-EYED TREE FROG** sleeps during the day, a semitransparent membrane covers its eyes. This additional **EYELID** lets in a small amount of light so it can tell if a predator approaches. When its eyes open, the flash of color will startle the predator so the frog can make a quick getaway.

I'm shielded

by an armor of SCALES.

But did you know that

my tongue . . .

A **PANGOLIN**'s long tongue helps it feed on insects underground. Its tongue can reach over 1 foot (30 cm) long. How does a pangolin fit such a long tongue in its mouth? It doesn't! The tongue is kept in **ITS CHEST**.

When challenged,
I won't step aside.
I'm ready to FIGHT
when I . . .

OPEN WIDE!

When two **SARCASTIC FRINGEHEADS** battle, they wrestle by pressing their open mouths against each other, as if they were kissing. The fish with the **BIGGER MOUTH** wins!

I catch my prey
with TALONS
that are strong.

But you might
not know . . .

THAT MY LEGS ARE LONG!

While known for their lethal talons, **OWLS** get much of their **STRENGTH** from their long legs. When they're roosting, their legs are covered by their feathers.

I might look cute
with my BILL and fur.
But hidden from sight . . .

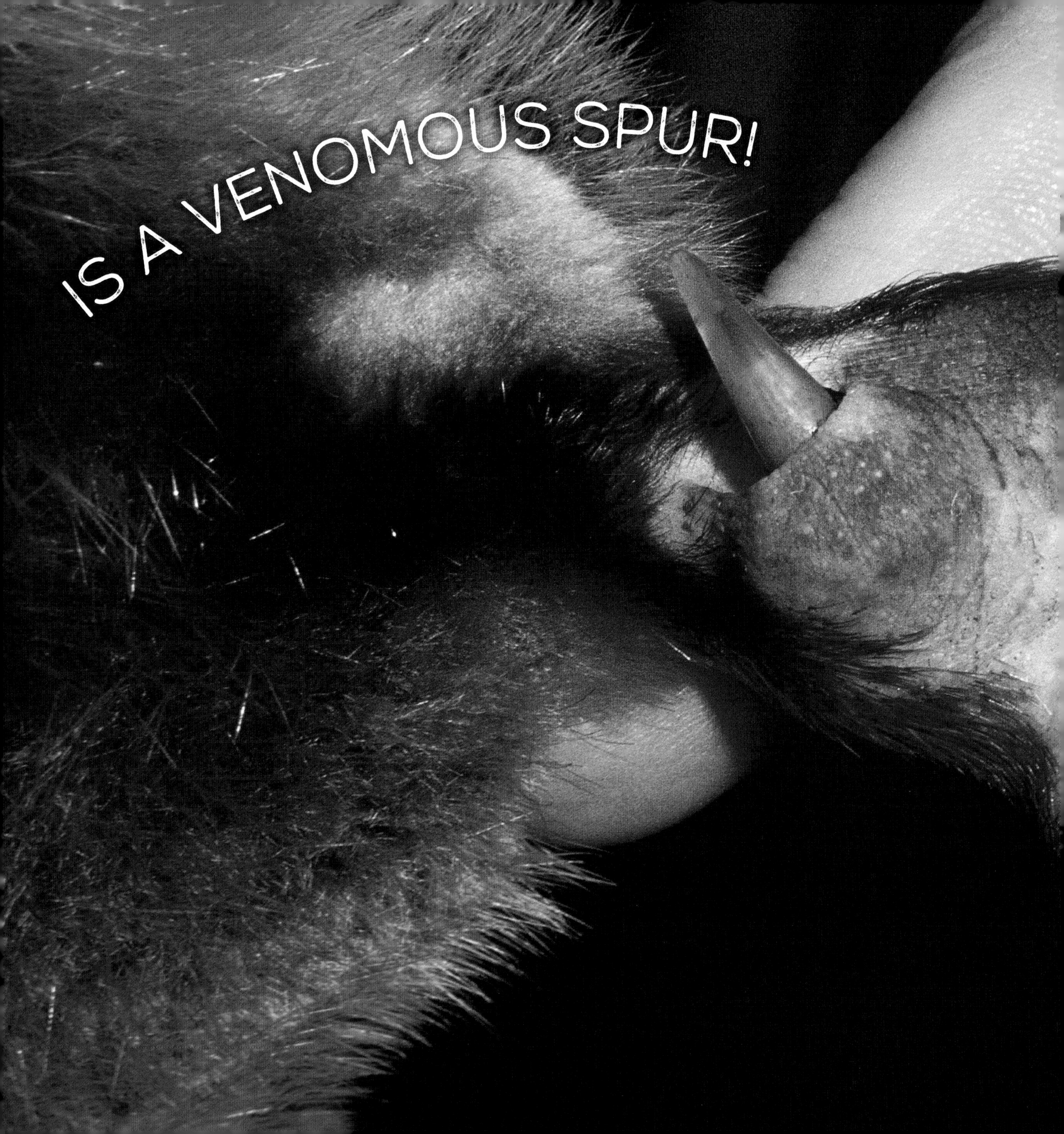

IS A VENOMOUS SPUR!

Male **PLATYPUSES** have a secret weapon: spurs on their hind feet that secrete **TOXIC VENOM**. They produce venom only during breeding season when they are competing for mates.

I EAT some foods
that you might like.
It could be why . . .

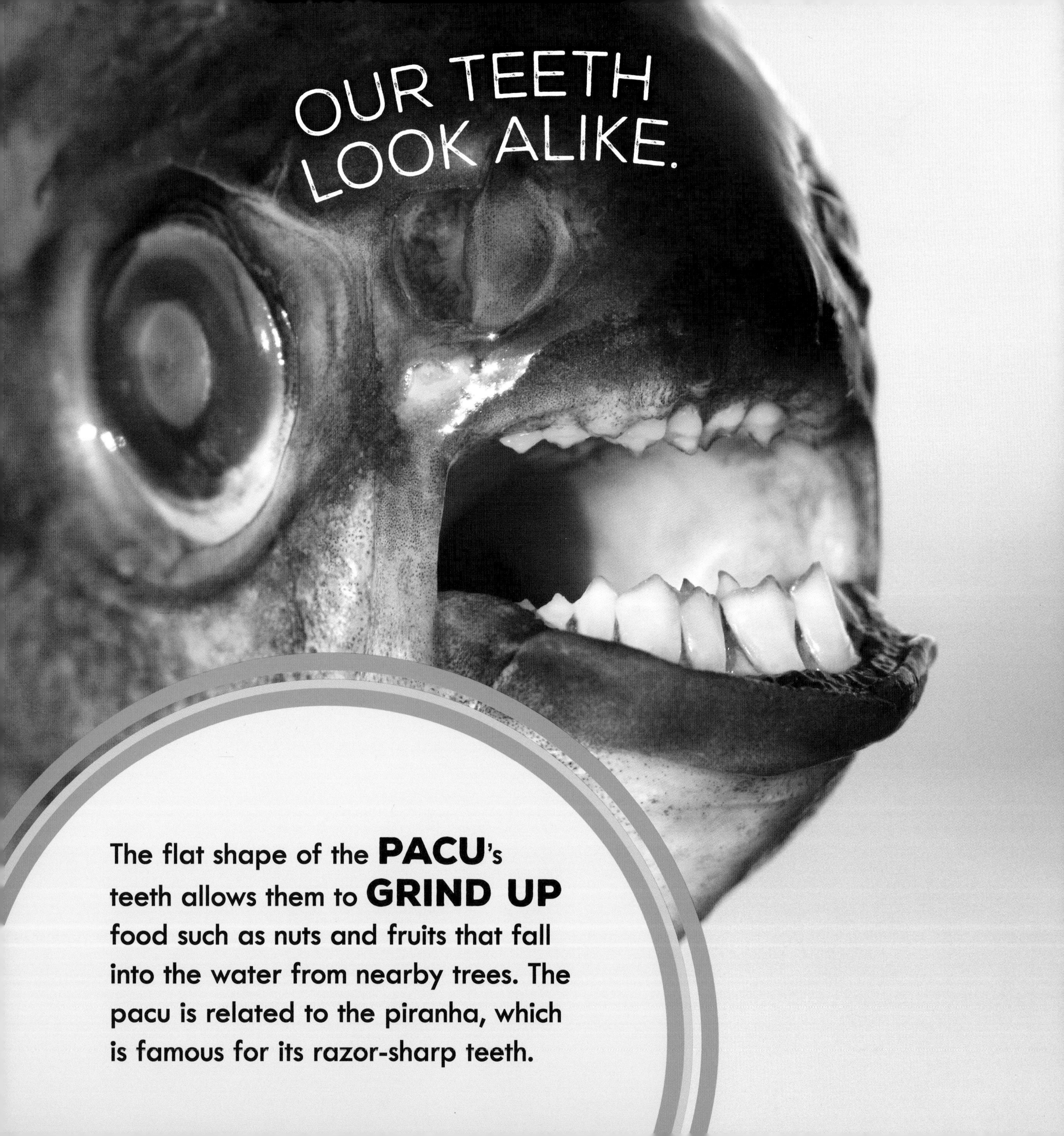

The flat shape of the **PACU**'s teeth allows them to **GRIND UP** food such as nuts and fruits that fall into the water from nearby trees. The pacu is related to the piranha, which is famous for its razor-sharp teeth.

If I run from a predator,
I might not win.
Instead, I'll ESCAPE
with my . . .

PARACHUTE SKIN!

The **COLUGO** has a flap of skin that connects all of its limbs. This gliding **MEMBRANE** allows it to "fly" between trees over 600 feet (183 m) apart.

We all have
HIDDEN FEATURES
that help us do
the things we do.

Tigers have webbed feet to help them swim.

What's something we might not see that makes you, YOU?

Animals have even more features than those included in this book. Different features help animals do different things.

Imagine a special feature that would help you if you had to

- CLIMB A TREE
- SWIM
- DIG HOLES
- FLY
- KEEP WARM
- SEE IN THE DARK

GLOSSARY

adaptation: a change in animals' appearance or behavior that helps them survive

feature: a notable part or characteristic of an animal

keratin: a hard protein that makes up hair, nails, horns, hoofs, and feathers

membrane: a thin layer that connects parts of an organism

parachute: a device that can slow an object falling through the air

semitransparent: a material that allows some light to pass through

spur: a sharp, hollow hook on a male platypus's back legs

talon: the claw of a bird of prey such as owls, hawks, and eagles

venomous: making a poisonous liquid that is usually passed on by a bite or sting

FURTHER READING

Books

Hanáčková, Pavla. *Why Penguins Don't Get Cold: How Animals Adapt to Their Surroundings.* London: Windmill Books, 2021.

Jenkins, Steve, and Robin Page. *Creature Features: Twenty-Five Animals Explain Why They Look the Way They Do.* Boston: Houghton Mifflin Harcourt, 2014.

Murphy, Julie. *Odd Bods: The World's Unusual Animals.* Minneapolis: Millbrook Press, 2021.

Piro, Radka. *Animal Adaptations: Unique Body Parts.* Prague: Albatros Media, 2021.

Websites

Idaho PBS: Animal Adaptation
https://sciencetrek.org/sciencetrek/topics/animal_adaptations/facts.cfm

National Geographic Society: Adaptation and Survival
https://www.nationalgeographic.org/article/adaptation-and-survival/

PBS: Animal Adaptation
https://ca.pbslearningmedia.org/resource/animal-adaptation-video/science-trek/

For Adeline and Grant

Millbrook Press™
An imprint of Lerner Publishing Group, Inc.
241 First Avenue North
Minneapolis, MN 55401 USA

For reading levels and more information, look up this title at www.lernerbooks.com.

Image credits: Dirk Ercken/Shutterstock, p. 1; anankkml/iStock/Getty Images, p. 2; TigerStocks/Shutterstock, p. 3; Paul Souders/Stone/Getty Images, p. 4; Ralph Eshelman/Shutterstock, p. 6; JasonOndreicka/Getty Images, p. 8; Ryan M. Bolton/Alamy Stock Photo, p. 10; David Brossard/Wikimedia Commons (CC BY-SA 2.0), p. 11; Mint Images/Getty Images, p. 12; Julian Gunther/Shutterstock, p. 13; Adriane Honerbrink, p. 14; Aluma Images/Getty Images, p. 16; pchoui/E+/Getty Images, p. 18; Doug Gimesy/Alamy Stock Photo, p. 19; slowmotiongli/Getty Images, p. 20; Afanasiev Andrii/Shutterstock, p. 22; CB2/ZOB/Supplied by WENN.com/Newscom, p. 24; Butterfly Hunter/Shutterstock, p. 25; Oliver Thompson-Holmes/Alamy Stock Photo, p. 26; AP Photo/Eric Risberg, p. 28; Sundraw Photography/Shutterstock, p. 30; bakabuka/Shutterstock, p. 32.

Front cover: Cathy Keifer/Shutterstock. Back cover: Paul Souders/Stone/Getty Images.

Designed by Emily Harris.
Main body text set in Tw Cen MT Std. Typeface provided by Monotype Typography.

Library of Congress Cataloging-in-Publication Data

Names: Park, Jane, 1972– author.
Title: Hidden creature features / by Jane Park.
Description: Minneapolis, MN : Millbrook Press, an imprint of Lerner Publishing Group, Inc., [2023] | Includes bibliographical references. | Audience: Ages 5–9 | Audience: Grades K–1 | Summary: “Animals have adaptations to help them survive. But some of those traits can be hard to spot. Rhyming text and eye-catching photos introduce eight creatures with hidden features” —Provided by publisher.
Identifiers: LCCN 2022020284 (print) | LCCN 2022020285 (ebook) | ISBN 9781728445670 (library binding) | ISBN 9781728485782 (ebook)
Subjects: LCSH: Animals—Adaptation—Juvenile literature.
Classification: LCC QH546 .P28 2023 (print) | LCC QH546 (ebook) | DDC 590—dc23/eng/20220607

LC record available at https://lccn.loc.gov/2022020284
LC ebook record available at https://lccn.loc.gov/2022020285

Manufactured in the United States of America
1-50264-49876-7/27/2022